Chromecast

10 Ultimate Tricks to Use Your Device Effectively

By Rick Moffet

Table of Contents

Introduction: What is Chromecast? 3

Chapter 1: Go Remoteless 6

Chapter 2: Mirror the Display 8

Chapter 3: Stream a Local File 11

Chapter 4: Improve the Quality 15

Chapter 5: Separate Audio and Video 17

Chapter 6: Your Best Photo Album 19

Chapter 7: Use It in a Hotel Room 22

Chapter 8: Play Games 25

Chapter 9: Visualize 27

Chapter 10: Guest Friendly 29

Conclusion 31

Introduction: What is Chromecast?

Chromecast is a small device that connects your phone or laptop to your TV. This amazing dongle developed by Google, helps you cast your desired type of entertainment on a high definition television, using mobile or web apps.

To explain it even better, think of your phone as a remote. You can relax on your sofa while streaming the photo album from your vacation on your TV, play your favorite song, or watch the last episode of that TV show you like on the BIG screen. And all this can be achieved in no time, without going through the trouble of the traditional transferring files while staring at that green 'loading sign', impatiently counting the remaining time– those irritating couple of minutes before you finally get to plug that flash drive off your computer.

Well, say farewell to all that effort, because Chromecast offers you something incredible. To enjoy the charms of this amazing gadget, all you need is a WiFi connection, a TV that has an HDMI port and a Google Chrome Browser, whether from your phone, computer, laptop or even tablet. What powers the device is a cable that is plugged into the Chromecast on one end, and the USB port of your TV on the other.

So far, there have only been two generations of chromecast:

1. Chromecast 1 is a device that is basically a stick, with an HDMI plug that is built in it.

2. Chromecast 2 is a disk-shaped gadget with an HDMI plug on a cable that is attached magnetically to the device. The advantage of this device is that unlike the first generation, Chromecast 2 doesn't require that much space, which is excellent for those whose HDMI ports are behind the TV which is against the wall. Besides the standard Chromecast 2, the second generation offers yet another practical device called Chromecast audio that streams only audio media.

Since the launch of the first chrome cast on July 24, 2013, over 20 million devices have been sold worldwide. With only $35 out of your wallet, this best-selling streaming gadget can take the entertainment in your home to a whole new level.

Although at the time of the Chromecast's launch there were only a few apps compatible with the device, today, this incredible tool allows you to enjoy thousands of different apps. Some of the most commonly used are:

- Chrome
- Netflix
- YouTube
- Hulu
- HBO
- Google Play Movies
- PIXLR
- Spotify
- Crackle
- Red Bull TV
- WatchESPN
- MediaBrowser
- Artkick
- Plex
- BubbleUPnP
- AllCast
- Crunchyroll
- iHeartRadio
- Pandora
- Twitch
- Ted
- And much, much more.

Chapter 1: Go Remoteless

How many times has it happened to you to turn your apartment upside down searching for your remote control? Hidden between the sofa cushions or even misplaced on the kitchen counter, remote controls are known to cause us trouble. But what if I tell you that your remote control is about to become nothing more than a dust collector? A useless tool that you will barely touch. Thanks to the power of the Chromecast device, you are one step behind to enjoying a relaxing afternoon without the controls of the remote.

With your Chromecast device always hooked up, the use of the remote control is really not necessary. From now on, you can let your mobile device be in charge and do all the work. If your Chromecast is powered on, your mobile phone can not only play media when the Chromecast is on, but it can also turn on your TV and even go to the Chromecast's HDMI input. This can all be achieved thanks to the HDMI-CEC technology that Chromecast supports. Most of the smart TVs also support this technology, but the feature may not be always enabled. If you cannot stream files directly from your phone, tablet or computer, then you'll have to go through your TV's settings and allow the HDMI-CEC to let the device you wish to manage the Chromecast, become your remote control. Go through your user manual to find that option. Depending on what brand is your TV, the name for the HDMI-CEC setting can be different. If you have a Samsung TV you can find it as 'Anynet +' if your TV is Toshiba you can find it as 'Regza Link', Sony TVs call this technology 'Bravia Sync', and if your smart TV is LG then the HDMI-CEC setting is called 'Simplink'.

Even though most of the TV brands call the HDMI-CEC 'link', 'net' or 'sync', I recommend you to search your manufacturer before you enable some setting that you are not completely sure about.

Once you manage to enable it, the real magic of being a Chromecast user can finally begin.

Chapter 2: Mirror the Display

If you have ever wished for a bigger screen, then you will find this Chromecast's advantage extremely helpful. Imagine sitting with your friends in your living room and surfing the net on your smartphone. Instead of reading and explaining to them what you have found online, Chromecast allows you to simply mirror the tab from your phone display to your TV so that everyone can see what you are looking at your screen. You can do the same with your tablet and computer. To enjoy this amazing feature that Chromecast offers you need to download the latest version of Google Chrome first. If you already have a Google Chrome browser, make sure to update it. If there is not a cast icon on the top right corner of your browser, download the *Google Cast Extension*. Then, follow these steps:

- Open the tab you wish to cast.
- Click the 'Cast' icon that is on the top right corner of your browser.
- Select your Chromecast device.
- When the icon turns blue, that means it has become active.
- Enjoy.

*Know that specific websites that use plug-ins such as Quicktime, VLC or Silverlight are not supported by

Chromecast, so if you wish to cast such site, the result will most likely be poor, lacking sound or picture.

For those of you that have phones running on Android, you are slightly more privileged than the iOS and Windows users, because, besides the casting tab option, you get to cast your entire screen without being annoyed by the lagging and bugging. If you are an Android user know that in order to cast your screen on the TV, your device must be running at least Android 4.4.2. To enable this option you have to:

- Download the Google Cast app.
- Make sure you are connected to the Chromecast's Wi-Fi network.
- Open the app.
- Look for the 'Cast Screen' button in the top left corner and tap it.
- Finally, select your Chromecast device.
- Enjoy.

*Know that Windows and iOS users have also the opportunity to cast entire screens. The difference between them and the devices running on Android is that Chromecast has labeled casting Windows and iOS screens as 'experimental', so the experience will not be so great. Fingers crossed Google will improve this feature in the nearest future.

To cast your entire screen and not only a single tab on Windows or iOS device, instead of clicking on 'Cast This Tab', choose 'Cast Entire Screen' from the Google Cast Extension.

Chapter 3: Stream a Local File

One of the main reasons why there are so many Chromecast users is the ability to stream local files without transferring them the traditional way. Instead of loading your flash drive with videos from your vacation, you can stream them on your TV through your Chromecast device.

But even though you may think of this magical device as a media transferer and player, it is actually more of an 'online tool' than it is a local player. Google's intentions with Chromecast are to make it an internet gadget that is tightly linked with Google Chrome. And it works perfectly; for $35 you get to stream media online with an impressive quality, but if you wish to enjoy those downloaded files or that amazing video you shot the other day, you may have to put some more effort.

There are a couple of different methods to stream the local files from your smartphone, tablet or computer.

Through Tab Casting. The simplest method of streaming local files is through your Google Chrome Browser. Again, the Google Cast Extension is obligatory. This method is quite simple; instead of casting a website, just use the CTRL-O command when your Google Chrome Browser is opened, and select the file you wish to cast. When a new tab with your file opens, click on the 'Cast Screen' button and select 'Cast This Tab'. Then, choose your Chromecast device, and enjoy. As easy as this procedure seems, there are quite a few limitations when it comes to streaming local files through the Tab Casting. For starters, not all files are supported (just the basic audio and video file types), and the highest resolution is 720p. However, the most appealing thing about this method is the fact that you don't have to have the video on full screen on your computer. Once you cast it on your TV, you can do something else on your computer and still have the file streaming through the Chromecast.

Through a 'Middleman'. If you have multiple short videos you want to stream, it is understandable why you would want to choose another way to do it, since the tab casting does not offer to create a playlist. Many choose the 'middleman' option when they simply install an app or a server that will help them stream local files from the computer to the TV (Chromecast). The most popular 'middlemen' are:

<u>Plex.</u> Plex is a media server that passes the media content to your Chromecast device. From your computer, you can stream pictures, music, and videos to your TV. There is also an option for queuing files, which will automatically create playlists. Here is how you can start enjoying the advantages of Plex:

- Sign up. In order to download Plex, it is required to have a Plex account. The account is free.
- Download Plex.
- Install the media server. Double-click the downloaded .exe file, click 'run', then 'install' and follow the steps.
- Start Plex. Once the installation is complete, double click on the Plex icon to start streaming files.   You will notice a '+' button. Click on it to add the desired file.
- Click 'Add Library'.
- Link the Plex server to your Plex account. In the settings menu, click 'connect' and then log in.

- After signing in, go to www.plex.tv again. Click 'launch' to get the web app.
- Now, click on the Chromecast icon on the top right corner and select your device.
- You can finally stream your files.

Videostream. Videostream is an app by Google Chrome that streams media to Chromecast. If you want a smooth video, in that case, I highly suggest Videostream, which, unlike Plex, it is extremely easy to install and very simple to use. The only downside to this app is that it only streams videos.

- Go to the Chrome Web Store and download Videostream for free.
- Click 'Add' and like all of the other Google Chrome extensions, Videostream will be installed instantly.
- The app will open automatically after the installation.
- Click 'Open' to select your desired file.
- A pop out window with Chromecast devices will be opened. Select yours.
- Enjoy.

Chapter 4: Improve the Quality

If you have bought your Chromecast mainly for Netflix, HBO or watching YouTube videos , you must agree that the streaming performance is actually pretty amazing. But if you often watch videos casting them via your Google Chrome Tabs, you must have experienced some stuttering and not-so-decent streaming quality. But what should you do? Should you just accept the fact that it was too good to be true for $35 or are you only clueless about technology and all the hidden 'rules' that can uplift your Sunday afternoon mood?

Sometimes the problem that's annoying us is a very simple matter. So before you go on and try to blame Google, you may want to detect the reason for the lagging. Most of the time, the reason for casting a poor quality video is nothing more than your wireless connection. Being an internet device, Chromecast mainly depends on the quality of your WiFi. Go to the Chromecast input to check for the signal strength. If it is low and you have Chromecast 1, you might want to try the HDMI extender that came with your Chromecast. Some TVs tend to block the WiFi signal and you literally need an extender that will protrude to get a good signal. Chromecast 2 has an HDMI extender built in for better performance.

If your signal is still low, you need to place your wireless router closer to your Chromecast dongle. This might make a serious change in streaming quality.

However, despite the obvious WiFi problem, many Chromecast users are faced with a different kind of stuttering. Those users are mainly those that have a not-so-speedy internet connection. To them, streaming online can be a real 'nightmare'. If you are one of those people, don't worry, there is still a way for you to put a stop to the buggy video you are trying to watch. In order to do so, you might want to check the resolution of the video you are streaming online. If you don't have the Google Chrome Cast Extension, please install it now. Then, click on the top right corner Cast button and select Options. Then a window with video quality settings will open. Adjust your resolution by selecting from Standard (480p), High (720p) and Extreme (720p high bitrate). Sometimes, it happens that the extreme high resolution is adjusted, and unless you have a super speedy connection, the streaming will be pretty bad. Although you will lower the quality of the picture resolution, the streaming will be smooth and you will finally get to enjoy what you were trying to watch all along. Pretty good for $35, don't you agree?

Chapter 5: Separate Audio and Video

How many times were you in the mood for enjoying your favorite TV show, but since your partner was taking his late nap, you were forced to miss all the action going on while watching TV with the lowest volume possible? Well, if you are Android or iOS user, you can say goodbye to not being able to feel the suspense. Not many Chromecast users know this, but there is a way to separate the sound and picture from your Chromecast. No, I am not talking about magic (although it does sound pretty magical), there is a magnificent app that allows you to stream your video via Chromecast on your TV, but keep the sound on your phone. The app is called LocalCast and as I said, it is only available to Android and iOS users.

To enjoy the charms of watching a loud movie late at night this is what you'll have to do:
- Go to Google Play and download the free LocalCast app.
- Beam the media via the app to your Chromecast device.
- Click 'Route Audio to Phone' on your 'Now Playing Screen'.
- Plug in your headphones or earphones.
- Watch your video on the big TV while the audio is playing on your phone.
- And everybody is happy.

If you go through the app description you will find that the LocalCast app is in beta and the performance may not be a high-class one, most of the users are more than satisfied with this amazing feature.

When it comes to separating audio and video, there is also a way not to route the audio on your phone, but to completely separate it from the video. You might think of your Chromecast as a video streamer only, but there is also a way to sort of make it a Chromecast Audio as well, without having to buy the device for additional $35. It is quite simple actually; all you need to do is to buy a cheap small adapter that will convert your HDMI to VGA and turn your Chromecast dongle into an audio adapter.

There is a lot of music available through Google Play Music, Pandora, Vevo, 8Tracks, Songza etc, as well as many radio apps such as TuneIn. Of course, with your Chromecast you can already cast your favorite songs to your TV, but with the trouble of buying a small and cheap adapter, you get the chance to connect your dongle directly to your stereo and enjoy the music.

Chapter 6: Your Best Photo Album

And while older people may tell you how ridiculous today's technology is, and how they miss the simple times of waiting for days to see how the taken picture has turned out, we, the people of today, greatly appreciate the ability to take as much as pictures as we like, then go through the selecting process and pick only the best ones. But how do you show off your photography skills? Do you print your photos, put them in frames and hang them on the wall? Or do you make a screensaver slide? It doesn't matter if you do the first or the latter, Chromecast offers you something amazing that will make you spend some more time with your camera.

Thanks to this incredible dongle's feature, you can now show your desired photos on your TV, without having to zoom-in so that everyone can see the image well, as well as create a screensaver photo album.

To play photos you will need to:

- Connect your device to the same network as the Chromecast.
- Download the Google Photos app.
- Open it.
- Tap the 'Cast' button on the top right corner.

- Select your Chromecast device.
- On your phone, tablet or computer, open the desired photo.
- Swipe the photos to change the displayed ones.
- When you are done, tap the Cast button again.
- Click 'Disconnect'.

*Your TV will be your best photo frame.

You may have noticed how your Chromecast plays a slideshow of some beautiful photos when you are not streaming any media. But, as wonderful as those photos are, I bet you all think how the screensaver would be much better if Chromecast played your own photo album. Luckily for you, this is something you can easily do. Here is how:

- Find the Chromecast app and install it on your phone.
- Tap the icon on the top right corner.
- Connect with your Google account.
- Exit the app.
- Download the Google Photos app.
- Open it.
- Create a new album and name it.
- Upload all the pictures you want to be part of the slideshow.

- Open the Chromecast app.
- Tap on 'devices'.
- Find your device and tap 'more'.
- Then, click on 'Backdrop Settings'.
- Tap 'Google Photos' on the screen.
- Find the 'Selected Albums' option and select your created album.
- Now, after the Chromecast slideshow, your own photo album will be displayed.

The ability to customize your Chromecast background does not only let you show off your vacations photos; using Backdrop you can also have weather information, read news and lifestyle (not in the UK), as well as enjoy satellite images.

Chapter 7: Use It in a Hotel Room

If you are a person who travels a lot for work, then you will greatly appreciate this tip. Imagine this scenario – after a long and busy day you are finally in your hotel room and you want to relax. But instead of paying an overcharge for renting a movie, you get to hook up with Netflix and catch up with the latest Game of Thrones episode. Wouldn't that be something? Although it seems like a complicated thing, getting your Chromecast to work when you are away from home, it can actually be easily achieved once you cover the essentials.

The first thing you must have in order to really make it work is an app called Conectify Hotspot. The hotel already has an internet connection, but this app will create a hotspot on your laptop that will allow you to connect to your Chromecast.

- Download Conectify Hotspot.
- Connect to the WiFi connection of the hotel.
- Open the app.
- Share the connection.
- Connect your Chromecast device to your new hotspot.

Although this sounds pretty simple, it isn't always. Most hotel's WiFi networks have AP/client isolation which is a good security for you as a guest, as it prevents other guests from accessing your files if you happened to had accidentally turned on sharing files. But, when it comes to Chromecast, the AP/client isolation is a real enemy.

Another problem is definitely the authentication splash screen where you need to accept the terms of service, enter your room number, register, and so on and so on, but with a hint of ingenuity even with this authentication, you can get Chromecast to work. The first thing you need to do (and something I strongly recommend) is to buy a portable router and an Ethernet cable. Most of the hotel rooms have Ethernet jacks, so there shouldn't be a problem. Plug your router into the Ethernet jack and turn it on. Then, connect to the router via your phone, tablet or laptop and configure your Chromecast device. Your router has already an IP or a Mac address so even if you have to accept terms of service with one device, the system will let any other device connected to the router (in your case your Chromecast device) to access the net. However, if the system has an authentication splash screen as well, you will have to use your creativity to get your Chromecast get to pass it.

The trick here is to use your Chromecast Mac address to complete the authentication process. Here is how:

- The MAC address is written on the back of the device, but since you have to have a super sharp vision to be able to read it, you need to find out the address the hard way.
- Plug your Chromecast in.
- Complete the setup to the step where you can see the MAC address.
- Turn it off.
- In the settings menu, change your device's address with the Chromecast's MAC address.
- Open the browser to finish the process of authentication.
- Change the address of your device with its original one; Reboot it.
- Authenticate it with its address.
- Fire the Chromecast.

This allows you to have both of your devices – your Chromecast and your phone, tablet or laptop connected to the network.

Even at first, it seems like much of a hassle to get your Chromecast to stream media in your hotel room, this process is actually not that hard once you learn what you have to do. Then, you can get comfortable in that giant bed, watching your favorite show.

Chapter 8: Play Games

Even though your Chromecast dongle isn't the first thing that comes to your mind when someone talks about gaming, this device offers a great way to kill boredom. Whether you have an hour or the whole day to kill, Chromecast have some great ways of entertaining you. The task is fairly simple. Visit the Chromecast store and download some fun games to play. Of course, have in mind that Chromecast is no Nintendo Wii or Sony Playstation so don't expect to play Battlefield or any other more complex game. But, if you're into card games, puzzle, trivia or strategy games, then playing them via Chromecast will be a true pleasure for you. These fun games are created for the whole family for a nice get-together night.

Once you have downloaded the game, look for the 'Cast' button. Its location varies for different games, as well as for different devices (iPad, Android Smartphone etc.).

The best part about playing games via Chromecast is that you don't have to do it alone. Many games offer the multiplayer option, so invite your friend and have some fun flying some angry birds. You are limited to games available for Chromecast only, of course, but if you both are Android users, then the list of fun games is really not short. So how to play? To enjoy a multiplayer game cast on your TV via Chromecast, first, you and your friend must both be connected to the same WiFi network. Then, the both of you need to connect to your Chromecast device. Once you have opened the game, your phones will become game controllers and the game will be displayed on your TV screen.

Some of the most popular games for Chromecast are:

- Monopoly
- Angry Birds
- Trivia bash
- Emoji Party
- Scrabble Blitz
- Yahtzee Blitz
- Risk
- Texas HoldEm
- Memory Cast
- Just Dance Now, and much more.

Chapter 9: Visualize

You have always envied those houses with fireplaces – those cozy places when the whole family snuggles and spends some quality time together, talking about their lives, laughing and enjoying their family nights, but you have never really felt that beauty. But who says that you need to have an actual fireplace to enjoy a comfy and relaxed time with your loved ones? Why don't you turn on your creativity and surprise your partner with a warm and quiet night for the two of you? If you have already spent $35 on the amazing dongle, anything is possible. Thanks to this super useful device, you get to wake up the romantic in you. Go to the Google Store and download a fireplace visualizer. Then, cast it on the TV via your Chromecast, just like you normally stream media. Besides the visual effect, you will also get an audio; a relaxing crackling that will get you into thinking you are sitting in front of an actual fireplace. Turn off the lights, light a few candles, open a nice bottle of wine and enjoy.

But the fireplace isn't the only visualizer that Chromecast offers. If the weather is cold and you cannot wait for the summer to arrive, why not imagine the summer is already there. Download a beach visualizer, hear the waves crash, seagulls chirp, fix a nice cocktail and enjoy. You can also do this with a snowfall visualizer. This is especially good for those of you who live in hot areas where it never snows.

Another visualizer that is very popular among Chromecast users is the aquarium.

Chapter 10: Guest Friendly

If it is your turn to entertain your friends and host a casual get together, then your Chromecast dongle can be more than just a handy device. You know how a social gathering goes; everyone is talking about some event and have some nice pictures to show the others. But wouldn't it be amazing if those pictures were cast on a big screen so that everyone can see them clearly, without having to pass the phone around?

Enabling a guest mode on your Chromecast will allow your guests to cast media even if they are not connected to your WiFi network.

- Open your Google Cast app
- Tap on 'Devices'
- Tap on the device card menu of your Chromecast
- Tap on 'Guest Mode'
- You can slide your guest mode to On and Off.

When it is on, your Chromecast will emit a WiFi beacon and when your guests launch the Google Cast App, the 'Cast' button will appear. Then, your Chromecast creates a 4 digit pin that can be found on your Chromecast backdrop. But how do your guests cast media?

- Open the Google Cast app
- Tap the 'Cast' button
- Tap on 'Nearby Devices'
- Wait until the devices is paired with the Chromecast
- If it fails, the guest will be asked to enter the 4 digit pin
- Start casting as you normally do.

*Chromecast guest mode works only for users whose devices are running Android 4.3 or higher, and iOS 7+.

*Know that iOS users must have their Bluetooth on, so they can start casting media.

*Guest mode for iOS is not available with the first generation of Chromecast.

Conclusion

Now that you have learned how powerful your Chromecast device is, there is only one thing left to do. Go on and let the power loose. Grab your smartphone or tablet, connect to the internet and get down to work. You would be surprised by what that little gadget can do.

If you thought you are not a technology person, wait to see the joy that Chromecast will bring to your life, and watch yourself transform into a real multimedia addict.

That's how good Chromecast is.

CPSIA information can be obtained
at www.ICGtesting.com
Printed in the USA
LVHW01s0141110618
580290LV00013B/302/P